YOUR KNOWLEDGE HAS VALUE

- We will publish your bachelor's and master's thesis, essays and papers

- Your own eBook and book - sold worldwide in all relevant shops

- Earn money with each sale

Upload your text at www.GRIN.com and publish for free

Imprint:

Copyright © 2016 GRIN Verlag, Open Publishing GmbH
Print and binding: Books on Demand GmbH, Norderstedt Germany
ISBN: 9783668334526

This book at GRIN:

http://www.grin.com/en/e-book/341711/major-land-uses-and-associated-negative-environmental-impact-an-empirical

Wuni Ibrahim Yahaya

Major Land Uses and Associated Negative Environmental Impact. An Empirical Investigation in the Bawku Municipality, Ghana

GRIN Publishing

Major Land Uses and their associated negative environmental impact in the Bawku Municipality: a focus on agricultural, residential, commercial uses and underlying issues

WUNI IBRAHIM YAHAYA

Department of Land Economy, Faculty of Built Environment, Kwame Nkrumah University of Science and Technology

Abstract: This journal article empirically explored the associated environmental impact of agricultural, residential and commercial land uses in the Bawku municipality. Questionnaires, researcher's observation, individual and group Interviews as well as focus group discussions were the survey instruments used to collect the primary data for the study from 150 respondents comprising farmers, home builders, hunters, shop keepers and stakeholders. The study established that most of the inhabitants 147 (98%) are not aware of existence of environmental permission for specialized size and quantum of land uses. The study also established that nearly all the inhabitants did not understand the negative environmental impact of their land use activities. The study also attributed the unhealthy environmental practices of the inhabitants to indiscipline of the land users. The study identified that environmental impact of the various land uses included threat to the ecosystem, depletion of the ozone layer, destruction of rare plant, tree and aquatic species and the a general extinction of some the rare plant and tree species. The reflective social and economic effect of the environmental destruction included threat to human survival, drought, harsh weather and climatic conditions and health related problems. The study in response to the causes of the unhealthy environmental practices accordingly presented a number of recommendations including creating awareness of the need to acquire environmental permit before commencing particular land uses and educating the inhabitants on the need to conserve the environment.

Keywords: Negative Environmental Impact, Major Land Uses, Empirical Investigation

Contents

1.0 INTRODUCTION

The environmental impact of land uses is becoming a force of global significance which was hitherto an issue of local environmental agenda. People try to satisfy their need from the land resources at the detriment of degrading environmental conditions. Countless of activities are been carried on land including residential, commercial, industrial, agricultural uses among other uses. It was projected by the UN-Habitat (2012) that 'human footprint has affected 83% of the global terrestrial land surface and has degraded about 60% of the ecosystem services in the past 50 years. Land use and land cover (LUCC) change has been the most visible indicator of the human footprint and the most important driver of loss of biodiversity and other forms of land degradation'

The natural environment supports the existence of man in these land uses. However, the activities of man relating to land uses have been known to be threatening the natural environment which as a consequence will threaten the survival of man afterwards. The various components of the natural environment are been destabilized from the uses to which man undertakes in the land. In response to this threat by man to the environment several campaign messages from both domestic and international organizations have been launched to create awareness of the need to protect the natural environment in our use of the land.

In the Bawku municipality, the natural environment (biodiversity) has been degraded in several ways in the quest of undertaking agriculture, residential property development and commercial land use activities. The trees are cleared to make way for houses, road constructions and other developmental projects without replacement. The birds are hunted and have resulted in the displacement of nearly all species of birds in the municipal. The rivers are poisoned for fishing purposes destroying nearly all species of the fishes but for very few areas. At worst because of the poverty level of the folks, charcoal farming has become a 'customarily approved' job in the municipal. In Towns like Garu and Tempane, the village folks have succeeded in cutting down nearly all the trees for 'Charcoal Burning" and for wood trading. There are also evidence of bush burning, unregulated refuse burning and chain-saw operations in the Municipal.

The environmental impacts of these unregulated environmental destructions for various means of survival and land uses have started attacking the communities in several ways. Drought and excessive rainless periods has struck the municipal in recent times. Most of the communities now facing threats of desert encroachment and during the time of the harmattan, the impact becomes unbearable. The sun heats the area without historical similarity.

It has become very obvious that, the inhabitants of the Bawku Municipality cannot marry the land uses with the natural environment and the signals are that if these rampant cutting of trees, unnecessary hunting of birds and poisoning of rivers are not regulated with immediate effects, the natural environment will soon become incapable of sustaining the lives of the people. The ordinary village folk do not understand the consequences of their actions and they see their actions as merely survival mechanisms.

This research journal article undertakes the empirical study of critical environmental impact of the major land uses and activities in the Bawku municipal. The study accordingly presents practical policy intervention at the local government level to curb the unhealthy environmental practices connected with the use of land.

2.0 RESEARCH METHODOLOGY AND MATERIALS

The journal made use of case study research design. A mixture of both quantitative and qualitative data was collected for the study. The quantitative data was collected using questionnaires and the qualitative data was collected using individual and group interviews and focus group discussions. The sampling techniques that were adopted for the study included purposive and snowball sampling. These were appropriate for the study because the sample frame could not be determined as a result of lack of record in that regard. The study included four communities and thus urban and rural characteristics were considered in their selection. Bawku Central (Sabongari) and Garu Central were the urban settings and Denugu as well as Yabrago constituted the rural settings. The study explored the impacts of agriculture on the environment from the perspective of the rural settings and therefore individual and group interviews were employed in that regard to collect the data. The residential and commercial uses of land and their associated environmental impact were investigated in the urban setting and involved the use of questionnaires and focus group discussions. Farmers, Hunters, Home Builders, shop owners and stakeholders constituted the respondents. A total of 150 respondents were studied comprising 10 hunters, 70 farmers, 50 home builders, 5 shop keepers and 15 other stakeholders. The data that was collected was aggregated and analyzed by categorizing issues into thematic areas, the use of percentages and frequency and cross tabulation. The method of reporting was basically descriptive.

3.0 PROFILING THE STUDY AREA

Bawku is a town and is the capital of the Bawku Municipal District, district in the Upper East region of north Ghana. It is located at 11°3′36″N0°14′24″W. Bawku has a Total Area of 1,275 km^2 (492 sq. mi). Bawku has a 2012 settlement population of 69,527 people (2012, estimates). Bawku is one of the Border Towns in Ghana. It is located in the extreme north-eastern corner of Ghana. Bawku is one of the municipal assemblies in Ghana.

The Bawku Municipal District is one of the nine (9) districts in the Upper East Region of north Ghana. It is made of the following constituencies: Bawku Central, Binduri and Pusiga. The district contains the following towns and villages. Of note are: Bawku, Pusiga, Garu, Denugu/Danvorga, Kongo, Zorsi, Tempane, Wuriyanga, Narango, Mognori (Gumbo), Widana, Yabrago, Missiga, Bugri-Bulpielse, Manga, Basyonde, Binduri Natinga, Kulugungu, Gozesi and Bugri.

The Kusaasis are the indigenous inhabitant population of the Bawku area. There are however large immigrant populations from other locations in northern and southern Ghana as well as from Burkina Faso, Ivory Coast, Togo, Niger and Nigeria. The district is characterized by agriculture, with Maize, Millet, Rice, tomatoes, soya beans and onions being amongst the main crops... The most dominant occupations of the inhabitants are farming, administrative works, teaching, food vending, Trading, Fishing, Artisanship, Auto mechanics among others

Bawku has a uniform temperature between 26°C and 33°C annually with an average annual temperature of 27°C. The speed of the wind in Bawku averages at 10km/h southwards and has an averages annual humidity of 77%. Bawku has a single rainfall maximum or regime. The sun is constantly hot throughout the year and always at its peak around May-June. Bawku is located in the tropical continental climate and in the Savannah vegetation zone. It has two seasons throughout the year; Thus Rainy season and Dry season. The soil components include loamy soil, sandy soil and clay soil

4.0 RESULTS AND DISCUSSION OF FINDINGS

4.1 Bio-Data of the Respondents

The study investigated a number of demographic features of the respondents on grounds that such characteristics of the respondents have influence of the environmental perceptions and understanding. The Bio-data covered the educational status of the respondents, the setting of the respondents, level of awareness on environmental permit, gender of the respondents and category uses to which they belonged. The analysis showed that majority of 150 respondents, 102 (68%) had no formal education and the remaining 48 (32%) were basically among the residential and commercial land users **(See Table 1)**. Most of the respondents were from the rural setting 95(63.33%) and the remaining 55(36.67%) were from the urban setting **(See Figure 2)**. It was also identified that 147(98%) of the respondents were not aware of the existence of environmental permit and this included a couple of the educated section of the respondents. The remaining 3(2%) who were aware of the environmental permits **(See Table 2)** were workers in the Municipal Assembly. This calls for the urgent need of creation of awareness in that regard. The study also included more males 80(53.33%) than women 70 (46.67%) **(See Figure 1)**. The study also included a survey of 80 (53.33%) agricultural land users, 50 (33.33%) residential land users and 20 (13.34%) commercial land users **(See Table 3)**. These were accordingly illustrated into tables and figures are below.

Table 1: Educational status of the respondents

Educational Status	No. of Respondents (%) N=150 (100%)
No Formal Education	102 (68%)
Basic Education	20 (13.33%)
Secondary Education	20 (13.33%)
Higher Education	8 (5.34%)

Source: Survey data, 2016

Table 2: Environmental Permit Awareness Status

Are you Aware?	No. of Respondents N=150 (100%)
Yes	3 (2%)
No	147(98%)

Source: Field Survey Data, 2016.

Table 3: Category of Land Users

Land Use	No. of Users N=150
Agriculture	80(53.33%)
Residential	50 (33.33%)
Commercial	20 (13.34%)

Source: Field Survey Data, 2016.

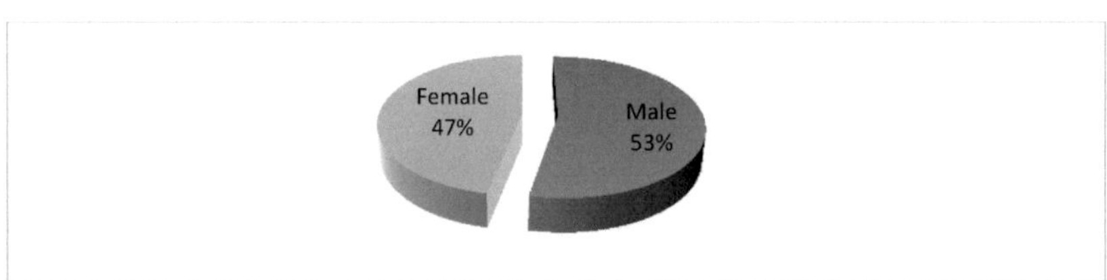

Figure 1: Gender of the Respondents

Source: Field Survey Data, 2016.

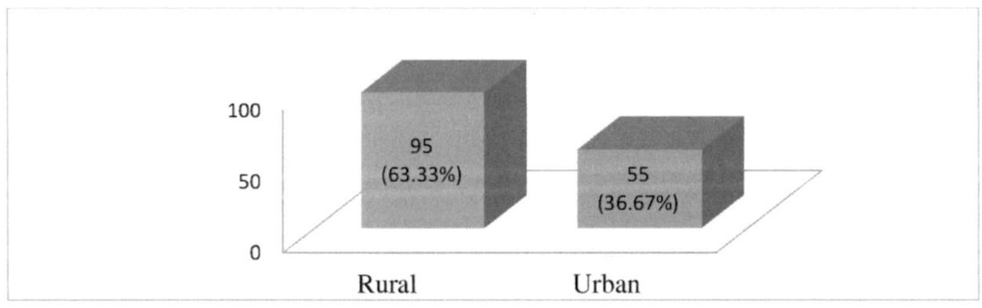

Figure 2: Location of the Respondents

Source: Survey data, 2016

4.2 RELATIONSHIP BETWEEN DEMOGRAPHIC FEATURES AND STATUS OF ENVIRONMENTAL PERMIT AWARENESS

4.2.1 *Location, Educational Status and Gender of the Respondents and Status of Environmental Permit Awareness*

The study further sought to carefully investigate whether or not the bio-data of the respondents had an influence on their level of awareness regarding the existence of environmental permit. The relationship between location of the respondent and that of environmental awareness status **(See Table 4)** reveals an uncorrelated pattern. All the respondents from the rural setting were not aware of the existence of environmental permit and just a few of those in the urban setting **3(5.45%)** are aware of the existence of the permit. This could be attributed to the fact that publicity and sensitization regarding the obligatory need of environmental permit for specific size and intensity of land uses has been duly neglected by the planning authorities. A similar pattern impression is also established from the relationship between educational status and awareness of the existence of environmental permit. All the respondents who had no formal education are not aware of the existence of the environmental permit **(See Table 5).** Only 3 of the educated were aware of the legal need of environmental permission for certain land uses. On the side of gender, only one woman from the urban setting is aware of the environmental permit and a single man from the same urban setting is aware of the environmental permit **(See Table 6)**. It can be established and concluded from the appraisal of the relationships that since the bio data had no great influence on the level of awareness, the publicity and the sensitization of the need for environmental permit has been neglected by the Municipal. This means the most of the inhabitants do not know the environmental consequences of the various land uses and this spells environmental doom for the municipal is not checked

Table 4: Location of the respondent and Environmental Permit Awareness Status

Location of Respondent	Are you Aware of Environmental Permit?		Total
	Yes	No	
Rural Setting	0 (0.00%)	95 (100.00%)	100%
Urban Setting	3(5.45%)	52 (94.55%)	100%

Source: Field Survey Data, 2016

Table 5: Education of the respondent and Environmental Permit Awareness Status

Educational Status the of Respondent	Are you Aware of Environmental Permit?		Total
	Yes	No	
No formal Education	0 (0.00%)	102 (100.00%)	100%
Basic Education	0 (0.00%)	20 (100.00%)	100%
Secondary Education	1 (5.00%)	19 (95.00%)	100%
Higher Education	2 (25.00%)	6 (75.00%)	100%

Source: Field Survey Data, 2016

Table 6: Gender of the respondent and Environmental Permit Awareness Status

Gender of Respondent	Are you Aware of Environmental Permit?		Total
	Yes	No	
Male	1(1.25%)	79 (98.75%)	100%
Female	1 (1.43%)	69 (98.57%)	100%

Source: Field Survey Data, 2016

4.3 Stakeholders' perspective of the Causes of the environmentally undisciplined Activities

Analysis of the data obtained from the stakeholders regarding the possible reasons for their uncaring attitude towards the environment, a number of motivations were identified to be drivers of such actions through the focus group discussions and the group interviews.

In the first place, most of the farmers undertook severe cutting down of trees without replacement because they perceived it to be a normal practice to clear trees and cultivate crops. This is a manifestation of utter ignorance. These groups are known not have any idea the reflective effects of cutting down trees without replace. A practical statement by one of the farmers in this regard was '……*the land is my own and so the trees. So if I want to farm and the trees are obstructing me, I will cut them down and continue….. (Apoko; a farmer from Denugu).* The stakeholders also identified that some of the motivations could be attributed to indiscipline because some of the farmers and hunters just burn bushes and trees in such of rats and honey (game). Bush burning is a great canker in the Bawku municipality and the cost has been loss of farms, animals and houses besides the environmental consequences.

Moreover, poverty was identified as one of the motivations. Bawku is indeed a low income municipality and which has been aggravated by 5 decades of chieftaincy disputes. All economic activities were brought to halt within this period and after 2009, when peace returned to the community poverty level of the inhabitants have heightened. A woman (named *Abugrikudu*) who was met burning charcoal from Yabrago (a rural setting in Bawku) sadly indicated '*i have five children and my husband is no more and I do not also have work so I feed my children through the income of charcoal farming.*

Unemployment was also identified as a contributory factor to environmental destruction. Most of the youth in the Bawku municipality are engaged in agriculture, fishing or illegal lumbering. They poison rivers to increase their gains from the fishes; they cut down very rare timber species for either charcoal burning or wood trading and clear smaller trees for agriculture purposes. When confronted they complain it is better to do that than to rob people. A graduate from one the polytechnics in Ghana disclosed that '*I completed poly in 2002 and I have been looking for a job and I am not getting; I do not also want to steal and so I have been surviving from cutting the trees for sale…… (Salifu; from Denugu).*

Ineffective and inefficient land use planning has also been identified as one of the factors increasing the environmental of the land uses. One can find incompatible land uses situated at one neighborhood in the Bawku Municipality for no any planning gain. These are signs of weak land use planning. Because of the non-enforcement of the land use planning regulations, the folks carry out any activity without restrictions or control. They capitalize on this and cut down trees and burn bushes as well as poison fishes.

4.4 THE ASSOCIATED ENVIRONMENTAL IMPACT OF THE VARIOUS LAND USES

The nucleus of the journal article is to move from conceptual analysis into an empirical investigation to clear unfold the environmental impact of the major land uses. The analyses presented in this journal are motivated from the findings from the field survey through focus group discussions from the stakeholders and observation as a native of the municipality.

4.4.1 THE ENVIRONMENTAL IMPACT FROM *AGRICULTURE LAND USES*

Picture of an over cultivated land with is nutrients depleted (Agricultural Land Use)

Source: Field Survey Photo, 2016

The unsustainable use of the land for agriculture is driving land degradation. This degradation of the land lead to the substantial loss of biodiversity in the municipal leading to threat plants and animals habitat, society and food security. The long term effect of this that land degradation impacts difficult or impossible to recover due to a neglect which lead to a cumulative addition to a long historical legacy of degradation

The clearing of the trees without corresponding replacement leads to deforestation, desertification and the consequent depletion of the ozone layer. Because the area has two seasons comprising a long dry season and a short rainy season, when the trees are fell, the action of the winds increases desertification of the affected area. Sometimes, the farmers engage the expertise of the chain-saw operators to cut down rare species of trees just to provide space for farming. This has particularly threatened a number of rare species. When the trees are also fell, the carbon dioxide (CO_2) content increases and when it attacks the ozone on a global scale, it depletes the layer.

In addition to the above, most farmers as part of their farming practices often burn the bush to hunt for games or undertake charcoal farming this expedite the depletion of the ozone layer as a result of the release of carbon mono-oxide. The chemical equation below best explains the effect of burning to the environment.

$$Carbon\ Mono\text{-}Oxide + Ozone \rightarrow Carbon\ (IV)\ Oxide + Oxygen$$

$$CO_{(s)} + O_{3(g)} \rightarrow CO_{2\ (g)} + O_{2(g)}$$

The carbon dioxide ($CO_{2\ (g)}$) produced from this reaction is not much absorbed by trees and plants in that particular area because they have all been cut down and the import of this is that the global carbon dioxide quantity increases automatically by that quantity. It might be argued that it is so little to be disastrous but *'little drops of water makes the might oceans'*

The unregulated use of fertilizer and agrochemicals for farming in the Bawku municipality releases harmful and persistent pollutants to the land, air and adjoining water bodies. The effect is loss of biodiversity and destruction of some plant species and some useful micro-organisms. This situation is aggravated by the practice of "bush burning before farming" which destroys a lot of plants and micro-organisms that sustain food production. The burning of bushes destroys terrestrial habitat. The biodiversity is threatened as most rare breeds of trees and plants are facing sharp extinction within the past few decades from agricultural practices in the area.

Another environmental impact from agricultural land uses is associated with depletion of nutrients, disruption of the biological cycles and soil erosion by continued cropping and land degradation with few or no inputs limits productivity over vast tropical and subtropical upland area.

In summary, the contemporary agricultural use practices in the muncipal are satisfying short-term increases in food production and security to the detriment of long-term losses in ecosystem and biodiversity.

4.4.2 THE ENVIRONMENTAL IMPACT FROM *RESIDENTIAL DEVELOPMENT*

In attempts to accommodate the inhabitants of the Bawku Municipality, several trees have been cut down to make way for such development. These trees are not necessarily replaced and the result is a reduction in the number and quantity of tree species. Most residential property developments in practice do not conform to zoning regulations because the home builders do not have planning, development and environmental permit for their buildings. They cut down trees anyhow and increasing the carbon dioxide content in the atmosphere.

In addition, the unregulated burning of refuse from the homes increases the carbon mono-oxide content in the atmosphere and this has consequently increased the depletion of the ozone layer.

4.4.3 THE ENVIRONMENTAL IMPACT OF *COMMERCIAL LAND USES*

The Bawku municipality is not well built up area and so a few commercial properties exist. Banks, Restaurants, Shops, Filling Station among others. The environmental impact from the commercial land uses is relatively less violent. However, except those commercial properties located in the Central Business Zone of the municipal, the rest of the buildings derive their space from the felling down of trees and consequently increases the carbon dioxide content in the atmosphere.

Another violent environmental impact for the commercial land uses includes the release of harmful and persistent pollutants, such as carbon mono-oxide from cars and vehicles. These among others threaten the quality of air and water in the community. There is the urgent need for strengthening of institutional and technical capacity in at the community level. This calls for the sustainable integration and effective implementation of existing controls at all levels in a community based-environmental management Programme.

5.0 THE REFLECTIVE SOCIAL AND ECONOMIC EFFECTS OF THE ENVIRON-
MENTAL DESTRUCTION

The cost of neglecting the environment from unregulated practices associated with major land uses in the Bawku Municipality comes in a number of ways.

There is the general threat to the survival of man and other livings things in the ecosystem. When the atmosphere becomes impure for human consumption, a number of diseases are bound to happen. The effects of the environmental degradation have been very direct on human health and such is increasingly well-understood. There has been an increase in the cases of cerebrospinal meningitis which has led to the loss of several lives and in the least paralyzed a sizeable number of people.

Desertification has been eating deep into the fabric of the municipal and it is a result of the dry nature of the lands in the municipal and extreme impact of the land degradation process. The combined effect has been very disastrous and strongly requires strengthening of the institutional and technical capacities of all stakeholders to address the menace..

There has also been extreme cases of severe drought for the past few decades and this could be traced to the excessive cutting down of trees. The result of these has been poor harvest and crops failure and the resulting increase in hunger and poverty.

One effect of the environmental destruction in Bawku is the increase in the impact of the sun-rays. The evidence of this happens around May-June of every year for the past 2 decades. During these times, the rooms and the earth because so heated that one could feel the heat on a mattress. Drinking water gets heated and the rooms become hotter and people tend to seek refuge in their courtyards in the evening.

In a nutshell, the social and economic problems rendered by the environmental impacts manifest in a number of ways including Changes in the atmospheric conditions, extensive modification of the earth's ecosystem, declines in biodiversity through the loss, modification, and fragmentation of habitats, degradation of soil, overexploitation of native species, many land use practices can degrade forest ecosystem conditions and loss of forest stock.

6.0 CONCLUSION AND RECOMMENDATION

It can be deduced from the findings that nearly all the inhabitants of the municipal are not aware of the existence and need for environmental permission for special types of land uses. The people in this regard do not really know the consequences of their actions regarding the environment and the result has been 'tragedy of the commons'. The municipal Assembly must be blamed for this neglect in the enforcement of planning permission acquisition before initiating a land use. In view of the violent and seemingly irreparable nature of the environ-mental effects, the study concluded that the law enforcement attitude of the relevant authori-ties are irreproachable and makes the following recommendations.

- o In the first place, the inhabitant should be educated on the need for environmental permit to be obtained before commencing particular land uses. Within the awareness creation, the people should be properly educated on the need to protect and conserve the environment. The message should be reasonably sustained to establish in the inex-tricable link between the environment and human survival. The planning authorities in the municipal assemblies shall be the appropriate agents in this regard. The people should be made to understand that there is the need for intergenerational conservation of the natural resources and hence the need to trade off short term individual benefits of a great future mutual and national benefits.

- o The planning authorities in the assembly in collaboration with the police should en-sure the enforcement of acquisition of planning, development and environmental permit before commencing any major land use. The permission acquisition process should be made less cumbersome and less expensive to encourage and motivate the folks to patronize. When this is done and sustained, land uses can be monitored at the earliest stage before it depletes the environmental architecture.

- o The planning authorities, assemblymen and the Ghana Forestry Department should institutes practical measures to stop the activities of the chain-saw operators, charcoal farming and rampant bush burning especially at the later part of the rainy season. This will eventually reduce the destruction the plants and trees species by these unscrupu-lous land users.

- o It is recommended that, the government of Ghana, stakeholders and capacity building organization should make sustainable and conscious effort to create non-farm jobs to absorb some the youth who are major culprit of the environmental destruction. And better still, best farming practices which encourages the interweaving of the agricultural activity with the natural environment should be employed and applied to ensure environmental conservation in agricultural practices.

o It is finally recommended that, strict enforceable punitive measures should be established to punish inhabitants who after the sensitization are still found guilty of any the unhealthy environmental practices. Justice must not be tempered with mercy in this regard. These are needed if the local government is prepared to start doing something about the environmental destruction.

Reference

United Nation's Department of Economic and Social Affairs (2012). *Sustainable land use for the 21st century.*

YOUR KNOWLEDGE HAS VALUE

- We will publish your bachelor's and
  master's thesis, essays and papers

- Your own eBook and book -
  sold worldwide in all relevant shops

- Earn money with each sale

Upload your text at www.GRIN.com
and publish for free